AF425430

Chemical Engineering for Students
Unleashing the Power of Chemical Reactions

currer

Chemical Engineering for Students Unleashing the Power of Chemical Reactions

Copyright © 2023 by currer

All rights reserved. No part of this book may be reproduced or transmitted in any form or by any means, electronic or mechanical, including photocopying, recording, or by any information storage and retrieval system, without permission in writing from the publisher.

This book is a work of fiction. Names, characters, places, and incidents either are the product of the author's imagination or are used fictitiously. Any resemblance to actual events, locales, persons, living or dead, is entirely coincidental.

The first edition was published in 2023

ISBN:
Published by:
Sunshine
1663 Liberty Drive
Hyderabad, IN 47403
www.Sunshinepublishers.com

This book is self-published using on-demand printing and publishing, which allows it to be printed and distributed globally

Table of content

Chapter 1: Introduction to Chemical Engineering 07

The Role of Chemical Engineering in Society

The Importance of Chemical Reactions in Engineering

Chapter 2: Fundamentals of Chemical Reactions 11

Types of Chemical Reactions

Stoichiometry and Reaction Balancing

Reaction Rates and Kinetics

Chapter 3: Thermodynamics in Chemical Engineering 17

The Laws of Thermodynamics

Enthalpy, Entropy, and Gibbs Free Energy

Thermodynamic Equilibrium and Spontaneity

Chapter 7: Multiphase Reactors 45

Gas-Liquid Reactors

Gas-Solid Reactors

Liquid-Liquid Reactors

Chapter 8: Safety in Chemical Reactions 51

Hazards and Risk Assessment

Safety Measures and Protocols

Emergency Response and Incident Management

Chapter 9: Industrial Applications of Chemical Reactions 57

Petrochemical Industry

Pharmaceutical Industry

Food and Beverage Industry

Environmental Engineering

Chapter 1: Introduction to Chemical Engineering

The Role of Chemical Engineering in Society

Chemical Engineering is a fascinating field that plays a crucial role in shaping the world we live in today. From the production of essential materials to solving environmental challenges, chemical engineers are at the forefront of creating sustainable solutions. In this subchapter, we will explore the significant contributions of chemical engineering in society.

One of the primary roles of chemical engineering is in the manufacturing industry. Chemical engineers are responsible for designing and optimizing processes that convert raw materials into valuable products. They work tirelessly to improve the efficiency of production methods, reduce waste generation, and minimize energy consumption. This not only benefits the companies but also contributes to the overall economic growth of a nation.

Moreover, chemical engineers are instrumental in developing new materials that revolutionize various sectors. They play a vital role in designing advanced polymers, composites, and nanomaterials with enhanced properties. These materials find applications in fields as diverse as aerospace, electronics, healthcare, and renewable energy. Chemical engineers are constantly pushing the boundaries of material science to create innovative solutions for a sustainable future.

Another crucial aspect of chemical engineering is its role in environmental protection. As the world faces pressing challenges such as climate change and pollution, chemical engineers are actively involved in developing cleaner technologies. They devise methods to reduce greenhouse gas emissions, treat wastewater, and manage hazardous waste safely. The goal is to find sustainable solutions that balance economic growth with environmental preservation.

In addition to their technical skills, chemical engineers also possess strong problem-solving and analytical abilities. They are trained to approach challenges from a systems perspective, considering the social, economic, and ethical implications of their work. This holistic approach enables chemical engineers to contribute to the betterment of society by addressing complex issues and finding practical solutions.

For students aspiring to pursue a career in chemical engineering, the opportunities are vast. The field offers a wide range of career paths, including research and development, process design, environmental consulting, and project management. With their unique skill set, chemical engineers have the potential to make a significant impact on society and contribute to the well-being of future generations.

In conclusion, chemical engineering plays a vital role in society by driving innovation, creating sustainable solutions, and addressing environmental challenges. As students, you have the opportunity to unleash the power of chemical reactions and make a positive impact on the world. By pursuing a career in chemical engineering, you can contribute to the betterment of society, shaping a brighter and more sustainable future for all.

The Importance of Chemical Reactions in Engineering

Chemical engineering is a fascinating field that combines the principles of chemistry, physics, and mathematics to design and develop processes that transform raw materials into useful products. At the heart of this discipline lies the concept of chemical reactions, which are vital for achieving desired outcomes in various engineering applications. In this subchapter, we will explore the importance of chemical reactions in engineering and how they unleash the power of innovation and progress.

Chemical reactions serve as the foundation for numerous processes in chemical engineering. Whether it is the production of pharmaceuticals, the development of sustainable energy sources, or the creation of advanced materials, chemical reactions are at the core of these advancements. Understanding and manipulating these reactions is crucial for engineers to optimize the efficiency, safety, and sustainability of these processes.

One key aspect of chemical reactions in engineering is their role in energy conversion. Engineers harness the power of reactions to convert one form of energy into another, such as converting chemical energy into electrical energy. For example, in the design of fuel cells, chemical reactions occur at the electrodes, enabling the conversion of chemical energy stored in fuels into electrical energy. This application has immense potential in the development of clean and efficient energy sources.

Moreover, chemical reactions enable engineers to synthesize valuable products. By carefully selecting reactants and controlling reaction conditions, engineers can transform raw materials into high-value chemicals, fuels, and materials. This has significant implications for industries such as pharmaceuticals, where chemical reactions play a crucial role in the synthesis of life-saving drugs.

Chemical reactions also play a pivotal role in waste treatment and environmental engineering. Engineers utilize various chemical reactions to remove pollutants from wastewater, reduce emissions from industrial processes, and remediate contaminated sites. Understanding the kinetics and thermodynamics of these reactions is vital to ensure effective and sustainable solutions for environmental challenges.

In conclusion, the importance of chemical reactions in engineering cannot be overstated. They are the driving force behind advancements in various fields, including energy, materials, pharmaceuticals, and environmental engineering. Chemical engineering students must develop a deep understanding of these reactions to unlock the full potential of their profession. By mastering the principles of chemical reactions, students can contribute to the development of sustainable and innovative solutions that address the pressing challenges of our time.

Chapter 2: Fundamentals of Chemical Reactions

Types of Chemical Reactions

In the exciting world of chemical engineering, understanding the different types of chemical reactions is of utmost importance. Chemical reactions are the backbone of this field, as they allow us to transform raw materials into valuable products that benefit society. This subchapter will explore the various types of chemical reactions that students studying chemical engineering need to comprehend in order to unleash the power of these reactions.

1.   Combustion   Reactions: Combustion reactions are commonly known as burning reactions. In these reactions, a substance combines with oxygen to produce heat and light. This type of reaction is essential in processes such as power generation, where fossil fuels are burned to generate electricity.

2.   Synthesis   Reactions: Synthesis reactions involve the combination of two or more substances to form a new compound. These reactions are commonly used in the pharmaceutical industry to create new drugs and in the production of various chemicals.

3.   Decomposition   Reactions: Decomposition reactions occur when a compound breaks down into simpler substances. These reactions are crucial in waste management and recycling processes, as they help break down complex compounds into their basic components.

4.   Displacement   Reactions: Displacement reactions involve the exchange of elements between two

compounds. These reactions are commonly used in metal extraction processes and are also important in the production of alloys.

5.   Acid-Base   Reactions: Acid-base reactions occur when an acid reacts with a base to produce water and a salt. These reactions are fundamental in chemical engineering, particularly in the production of fertilizers, pharmaceuticals, and various industrial chemicals.

6.   Redox   Reactions: Redox reactions involve the transfer of electrons between species. These reactions play a crucial role in energy production, such as in fuel cells and batteries, as well as in the purification of metals.

Understanding the types of chemical reactions allows chemical engineering students to design and optimize processes, select appropriate catalysts, and predict the behavior of reactions under various conditions. By mastering these reactions, students can harness their power to develop innovative solutions to real-world challenges.

Throughout this book, we will delve deeper into each type of chemical reaction, exploring their principles, mechanisms, and applications in the field of chemical engineering. By the end of this journey, you will be equipped with the knowledge and tools necessary to unravel the mysteries of chemical reactions and contribute to the advancement of this fascinating discipline.

So, let us embark on this exciting exploration of the various types of chemical reactions and unlock the potential they hold for the world of chemical engineering.

Stoichiometry and Reaction Balancing

Understanding stoichiometry and balancing chemical reactions is crucial for students pursuing a degree in chemical engineering. In this subchapter, we will delve into the fundamental concepts of stoichiometry and provide a comprehensive guide on how to balance chemical reactions effectively.

Stoichiometry is the branch of chemistry that deals with the quantitative relationships between reactants and products in a chemical reaction. It allows us to determine the amount of reactants needed, the amount of products formed, and the ratio between them. This knowledge is indispensable for chemical engineers as they strive to optimize reaction conditions and design efficient processes.

One of the first steps in stoichiometry is balancing chemical reactions. A balanced equation ensures that the law of conservation of mass is upheld, meaning that the total mass of the reactants equals the total mass of the products. Balancing equations involves adjusting coefficients in front of each compound or element to achieve this balance. Through this process, students will learn how to interpret and manipulate chemical equations, gaining a deeper understanding of reaction mechanisms.

We will explore various techniques for balancing equations, including the inspection method and algebraic method. The inspection method involves visually examining the equation and adjusting coefficients accordingly. While this method works for simple reactions, more complex reactions require a systematic approach. The algebraic method involves setting up a system of equations based on the conservation of mass, followed by solving for the unknown coefficients.

Additionally, we will cover the concept of limiting reactants, which plays a critical role in determining the extent of a reaction. Understanding the concept of limiting reactants allows chemical engineers to optimize reaction conditions and maximize product yield.

Throughout this subchapter, we will provide practical examples and step-by-step explanations to ensure students grasp the concepts and can apply them to real-world situations. We will also introduce software tools and calculators that aid in stoichiometry calculations, enabling students to streamline their work and improve their efficiency.

By mastering stoichiometry and reaction balancing, students will acquire a solid foundation in chemical engineering and be better equipped to tackle complex reactions and design innovative processes. This subchapter will provide a strong framework for their future studies and professional endeavors in the field of chemical engineering.

Reaction Rates and Kinetics

In the exciting world of Chemical Engineering, understanding the rates at which chemical reactions occur is of utmost importance. Welcome to the subchapter on Reaction Rates and Kinetics, where we will delve into the fascinating realm of how reactions progress and the factors that influence their speed.

First, let's define reaction rates. Reaction rates refer to the rate at which reactants are converted into products over a certain period of time. The study of reaction rates is crucial for chemical engineers as it allows them to design and optimize processes, control reactions, and maximize efficiency.

One of the fundamental concepts in reaction rates is the concept of kinetics. Kinetics is the branch of chemistry that focuses on studying the factors that affect the rate of chemical reactions. It helps us understand the mechanisms behind reactions, the role of catalysts, and the impact of temperature, concentration, and surface area on reaction rates.

Temperature plays a crucial role in reaction rates. As the temperature increases, the kinetic energy of the reactant molecules also increases, leading to more frequent and energetic collisions. This results in a higher reaction rate. Understanding the temperature dependence of reaction rates is essential for controlling and optimizing processes in chemical engineering.

Another factor that affects reaction rates is concentration. In general, as the concentration of reactants increases, the rate of reaction also increases. This is because a higher concentration provides a greater number of reactant particles, leading to more collisions and a higher likelihood of successful reactions.

Surface area is another important factor in reaction rates, especially for heterogeneous reactions. Increasing the surface area of solid reactants by grinding or using catalysts can significantly enhance the reaction rate. This is because a larger surface area provides more active sites for collisions to occur, increasing the chances of successful reactions.

Catalysts are substances that speed up reactions without being consumed in the process. They work by lowering the activation energy required for a reaction to occur. Understanding the role of catalysts in reaction rates is crucial for chemical engineers, as they can greatly enhance the efficiency and economic viability of industrial processes.

In conclusion, reaction rates and kinetics are vital concepts in the field of Chemical Engineering. Understanding the factors that influence reaction rates, such as temperature, concentration, surface area, and catalysts, allows engineers to design and optimize processes, control reactions, and maximize efficiency. By delving into the world of reaction rates and kinetics, students in the field of Chemical Engineering can unleash the power of chemical reactions and contribute to the advancement of this exciting discipline.

Chapter 3: Thermodynamics in Chemical Engineering

The Laws of Thermodynamics

Thermodynamics is a fundamental branch of science that deals with the study of energy and its transformations. In the field of chemical engineering, understanding the laws of thermodynamics is crucial as it forms the basis for analyzing and designing chemical processes. In this subchapter, we will delve into the laws of thermodynamics and explore their implications in the context of chemical engineering.

The first law of thermodynamics, also known as the law of energy conservation, states that energy cannot be created or destroyed in an isolated system. It can only be transformed from one form to another. This law is essential in chemical engineering as it helps us understand the energy balance in processes such as heat transfer, chemical reactions, and phase changes. By applying the first law, we can determine the amount of energy required or released during these processes, enabling us to optimize and control them effectively.

The second law of thermodynamics introduces the concept of entropy, which quantifies the degree of disorder or randomness in a system. It states that in any spontaneous process, the total entropy of an isolated system always increases. This law is particularly significant in chemical engineering as it guides us in understanding the direction and efficiency of chemical reactions and heat transfer processes. By considering entropy, we can determine the feasibility and limitations of chemical processes and develop strategies to maximize their efficiency.

The third law of thermodynamics states that the entropy of a perfectly crystalline substance approaches zero as the temperature approaches absolute zero. While this law might seem less relevant to chemical engineering students at first, it plays a significant role in understanding the behavior of materials at extremely low temperatures. It is particularly relevant in the fields of cryogenics and materials science, where the properties of materials at near-zero temperatures are crucial for various applications.

Understanding and applying the laws of thermodynamics is essential for chemical engineering students as it provides a solid foundation for analyzing and designing chemical processes. By grasping the concepts of energy conservation, entropy, and crystalline substances, students can optimize the efficiency and feasibility of chemical reactions, heat transfer processes, and material behavior. The laws of thermodynamics serve as a powerful tool for unlocking the potential of chemical reactions and harnessing the power of energy transformations.

Enthalpy, Entropy, and Gibbs Free Energy

Understanding the concepts of enthalpy, entropy, and Gibbs free energy is essential for any student studying chemical engineering. These concepts play a crucial role in understanding the behavior and transformation of chemical substances and reactions.

Enthalpy, denoted by the symbol H, is a measure of the heat energy absorbed or released during a chemical reaction. It is a thermodynamic property used to quantify the energy content of a system. Enthalpy change, ΔH, is the difference in enthalpy between the reactants and products of a reaction. By studying the enthalpy change, we can determine whether a reaction is exothermic (releases heat) or endothermic (absorbs heat). Understanding enthalpy allows chemical engineers to design and optimize processes that involve heat transfer, such as distillation or heat exchangers.

Entropy, denoted by the symbol S, is a measure of the degree of disorder or randomness in a system. It is another important thermodynamic property that helps us understand the spontaneity and directionality of chemical reactions. Entropy change, ΔS, provides insights into the changes in the molecular arrangement and energy dispersal during a reaction. By analyzing the entropy change, chemical engineers can predict whether a reaction is likely to occur or not. The second law of thermodynamics states that the entropy of an isolated system always increases or remains constant, highlighting the importance of entropy in understanding the behavior of chemical systems.

Gibbs free energy, denoted by the symbol G, combines both enthalpy and entropy to determine the spontaneity of a chemical reaction. Gibbs free energy change, ΔG, is a measure of the energy available to do useful work at constant temperature and pressure. A negative ΔG indicates a spontaneous reaction, while a positive ΔG indicates a non-spontaneous reaction. Chemical engineers use the concept of Gibbs free energy to assess the feasibility and efficiency of chemical processes. By manipulating the temperature, pressure, and composition, they can optimize reactions to minimize energy consumption and maximize the desired product.

In summary, enthalpy, entropy, and Gibbs free energy are fundamental concepts in chemical engineering. They provide the tools necessary to analyze and predict the behavior of chemical reactions. By understanding these concepts, students can design and optimize processes to unleash the power of chemical reactions and contribute to the advancement of the field of chemical engineering.

Thermodynamic Equilibrium and Spontaneity

In the fascinating world of chemical engineering, understanding the concepts of thermodynamic equilibrium and spontaneity is crucial. These concepts lay the foundation for predicting and controlling the outcomes of various chemical reactions, enabling engineers to design efficient processes and develop innovative solutions. In this subchapter, we will delve into the intricacies of thermodynamic equilibrium and spontaneity, unraveling their significance in the realm of chemical engineering.

Thermodynamic equilibrium is a state in which a system reaches a stable condition with no net change occurring over time. It is the point where the rates of forward and reverse reactions become equal, resulting in a balance between reactants and products. This equilibrium can be achieved under various conditions, such as constant temperature, pressure, and composition. By studying thermodynamic equilibrium, chemical engineers can determine the optimal conditions for a reaction to occur and design efficient reactors accordingly.

Spontaneity, on the other hand, refers to the natural tendency of a reaction to occur without any external influence. A spontaneous reaction is one that proceeds on its own, releasing energy and moving towards equilibrium. Chemical engineers must comprehend the factors that govern spontaneity, such as enthalpy, entropy, and Gibbs free energy. These thermodynamic properties help us understand the energy changes and disorder in a system, allowing engineers to predict the direction and feasibility of a reaction.

In this subchapter, we will explore the relationship between thermodynamic equilibrium and spontaneity, emphasizing their significance in chemical engineering. We will discuss the laws of thermodynamics, including the first and second laws, which provide fundamental principles for understanding energy transfer, work, and entropy changes. Additionally, we will delve into the concept of Gibbs free energy, a powerful tool for predicting the spontaneity of reactions.

Furthermore, we will examine real-world examples and case studies showcasing the application of thermodynamic equilibrium and spontaneity in chemical engineering. From designing efficient chemical reactors to optimizing energy consumption, these concepts play a vital role in various industrial processes, including petroleum refining, pharmaceutical manufacturing, and environmental engineering.

By mastering the concepts of thermodynamic equilibrium and spontaneity, students of chemical engineering can unlock the true power of chemical reactions. This subchapter aims to provide a comprehensive understanding of these concepts, equipping students with the knowledge and skills necessary to tackle complex engineering challenges in the future.

Chapter 4: Chemical Reactors

Introduction to Chemical Reactors

Chemical reactions form the backbone of the field of chemical engineering. Understanding the principles and mechanisms behind these reactions is crucial for students pursuing a career in chemical engineering. This subchapter, titled "Introduction to Chemical Reactors," aims to provide students with a comprehensive overview of the fundamentals of chemical reactors.

Chemical reactors are vessels or systems in which chemical reactions take place. These reactions involve the transformation of reactants into desired products through various processes such as combustion, synthesis, and decomposition. Chemical engineers design and optimize reactors to ensure efficient and safe production of desired products on an industrial scale.

In this subchapter, we will delve into the key concepts and principles that govern chemical reactors. We will explore different types of reactors, including batch reactors, continuous-flow reactors, and catalytic reactors, and discuss their applications in various industries.

Furthermore, we will study the factors that influence the performance of chemical reactors, such as reaction kinetics, heat transfer, mass transfer, and reactor design. Understanding these factors is essential for optimizing reaction conditions, maximizing yields, and minimizing undesirable by-products.

The subchapter will also introduce students to the different types of reactors used in chemical engineering, such as stirred tank reactors, tubular reactors, and fluidized bed reactors. We will discuss their advantages, limitations, and the selection criteria for choosing the most appropriate reactor for a given reaction.

Safety is of paramount importance in chemical engineering, and this subchapter will emphasize the importance of process safety in chemical reactors. Students will learn about potential hazards, such as runaway reactions, and the measures taken to prevent accidents and ensure the well-being of both workers and the environment.

Throughout the subchapter, real-life examples and case studies will be provided to illustrate the application of chemical reactors in various industries, such as petrochemicals, pharmaceuticals, and food processing. These examples will help students connect theoretical concepts with practical applications, enhancing their understanding of the subject matter.

By the end of this subchapter, students will have a solid foundation in the principles of chemical reactors. They will be equipped with the knowledge to analyze and design chemical reactors, optimize reaction conditions, and ensure the safe and efficient production of chemicals in various industrial settings.

"Introduction to Chemical Reactors" is an essential starting point for students aspiring to unleash the power of chemical reactions and embark on a successful career in chemical engineering.

Batch Reactors

In the world of chemical engineering, batch reactors play a vital role in the transformation of raw materials into valuable products. This subchapter will explore the basics of batch reactors, their operation, and their significance in chemical processes.

Batch reactors are vessels used to carry out chemical reactions in discrete batches. Unlike continuous reactors, which operate continuously with a steady flow of reactants and products, batch reactors handle a fixed quantity of reactants at a time. This makes them ideal for small-scale production, research, and development purposes.

The operation of a batch reactor involves several stages, including charging, reaction, and discharging. In the charging stage, the reactants are carefully measured and loaded into the reactor. Once the reactor is sealed, the reaction stage begins, where the reactants undergo chemical transformations under controlled conditions, such as temperature, pressure, and stirring speed. This allows engineers to optimize the reaction parameters and achieve the desired product. Finally, in the discharging stage, the product is collected, and the reactor is prepared for the next batch.

Batch reactors offer several advantages for chemical engineering students and professionals. Firstly, they provide better control over the reaction process, as the reactants can be adjusted and monitored throughout the reaction. This allows for greater flexibility in terms of reaction time, temperature, and reactant concentration.

Furthermore, batch reactors offer a safer environment for experimentation and process optimization. Since the reactants are contained within a closed vessel, the risk of accidental spills or leaks is significantly reduced. This allows students to explore various reaction conditions without compromising their safety.

Batch reactors also enable the study of reaction kinetics and the determination of reaction rates. By analyzing the changes in reactant concentrations over time, students can gain insights into the reaction mechanisms and optimize reaction conditions for improved product yield and selectivity.

In addition, batch reactors are widely used in the pharmaceutical industry for drug synthesis and production. The ability to carry out reactions in discrete batches allows for better control over the quality and purity of the final product, ensuring the safety and efficacy of medications.

In conclusion, batch reactors are an essential tool in the field of chemical engineering. Their versatility, control, and safety features make them ideal for students to learn and explore the power of chemical reactions. Understanding batch reactors and their operation is crucial for aspiring chemical engineers, as they are widely employed in research, development, and production processes across various industries.

Continuous Flow Reactors

In the world of Chemical Engineering, continuous flow reactors play a vital role in the efficient and controlled processing of chemical reactions. Understanding the principles and applications of these reactors is essential for students pursuing a career in this field. This subchapter aims to provide a comprehensive overview of continuous flow reactors, shedding light on their types, advantages, and applications.

Continuous flow reactors, also known as tubular reactors, are designed to handle reactions that occur continuously over time. Unlike batch reactors where the reactants are added and removed in discrete batches, continuous flow reactors allow for a continuous flow of reactants and products. This continuous operation offers several advantages, including improved control over reaction conditions, enhanced safety, and increased productivity.

There are various types of continuous flow reactors, each with its own unique characteristics and applications. The most commonly used types include plug flow reactors (PFR) and mixed flow reactors (MFR). In a PFR, reactants flow through a long tube, experiencing minimal mixing as they proceed. This type of reactor is suitable for reactions that require precise control over residence time. On the other hand, MFRs provide better mixing of reactants, making them suitable for reactions that require homogeneous conditions.

The advantages of continuous flow reactors extend beyond improved control and safety. They also offer enhanced heat transfer capabilities, allowing for better temperature control during reactions. Additionally, continuous flow reactors offer the ability to easily scale up production, making them ideal for large-scale industrial processes.

The applications of continuous flow reactors are widespread in the chemical industry. They are commonly used in the production of pharmaceuticals, where precise control over reactions is crucial. Continuous flow reactors are also employed in the petrochemical industry for processes such as catalytic cracking and reforming. Moreover, they find applications in the production of specialty chemicals, polymers, and fine chemicals.

In conclusion, continuous flow reactors are an integral part of Chemical Engineering, offering numerous advantages in terms of control, safety, and productivity. Understanding the principles and applications of these reactors is essential for students pursuing a career in this field. Whether it is in pharmaceuticals, petrochemicals, or other chemical industries, continuous flow reactors play a crucial role in unleashing the power of chemical reactions.

Reactor Design and Optimization

In the world of chemical engineering, reactors play a crucial role in transforming raw materials into valuable products through chemical reactions. Reactor design involves a careful analysis of various factors to ensure optimal performance, efficiency, and safety. This subchapter will explore the fundamentals of reactor design and the importance of optimization in the field of chemical engineering.

The design of a reactor involves the selection of the appropriate type of reactor, such as batch, continuous, or semi-batch, based on the specific requirements of the reaction. Each reactor type has its own advantages and limitations, and understanding these factors is essential for successful reactor design.

One of the primary considerations in reactor design is the reaction kinetics, which provides insight into the rate at which the reaction occurs. By studying reaction kinetics, students can determine the optimal conditions, such as temperature, pressure, and catalyst usage, for a specific reaction. This information is crucial for achieving high conversion rates and selectivity, while minimizing unwanted side reactions.

Furthermore, reactor design takes into account the heat and mass transfer requirements. Heat transfer is crucial to maintain the reaction at the desired temperature, while mass transfer ensures efficient mixing of reactants and products throughout the reactor. Students will learn about various heat transfer mechanisms, such as conduction, convection, and radiation, and explore mass transfer concepts like diffusion and dispersion.

Optimization is a key aspect of reactor design, aiming to improve the overall performance and efficiency of the system. By applying optimization techniques, students can identify the optimal reactor design parameters that maximize product yield, minimize energy consumption, and reduce environmental impact. This involves mathematical modeling, simulation, and analysis of the reactor system to determine the optimal operating conditions.

Moreover, safety considerations are of utmost importance in reactor design and optimization. Students will be introduced to various safety measures and protocols to ensure the prevention of accidents, such as the use of appropriate materials, proper ventilation, and emergency shutdown systems. Understanding the potential hazards associated with chemical reactions and how to mitigate them is vital for every chemical engineer.

In conclusion, the subchapter on Reactor Design and Optimization provides students with the fundamental knowledge required to design and optimize reactors in the field of chemical engineering. By understanding the principles of reactor design, reaction kinetics, heat and mass transfer, optimization techniques, and safety considerations, students will be equipped to unleash the power of chemical reactions and contribute to the development of sustainable and efficient chemical processes.

Chapter 5: Reaction Engineering Principles

Reactor Performance Metrics

In the world of chemical engineering, reactors play a crucial role in unleashing the power of chemical reactions. These reactors are designed to convert raw materials into valuable products through various chemical processes. However, assessing the performance of a reactor is essential to ensure its efficiency and effectiveness. This subchapter will delve into reactor performance metrics, providing students with a comprehensive understanding of how to evaluate the performance of these vital systems.

One key metric used to assess reactor performance is conversion efficiency. Conversion refers to the extent to which the reactants are converted into desired products. It is crucial to achieve high conversion rates as it directly impacts the overall yield and profitability of the process. Students will learn about the different types of conversion, such as single-pass conversion and overall conversion, and the factors that influence them.

Another important performance metric is selectivity. Selectivity measures the extent to which a reactor produces the desired product relative to undesired by-products or side reactions. Achieving high selectivity is critical to maximizing the yield of the desired product and minimizing waste. Students will explore the factors that influence selectivity, such as reactant concentration, temperature, and catalyst choice.

In addition to conversion and selectivity, students will also learn about reactor efficiency. Reactor efficiency measures how effectively a reactor utilizes the reactants and converts them into products.

Understanding reactor efficiency is crucial for optimizing the reaction conditions and designing reactors that minimize energy consumption and maximize output. This subchapter will cover both thermal efficiency and overall reactor efficiency, highlighting the significance of each in assessing performance.

Furthermore, this subchapter will introduce students to reactor design parameters that impact performance metrics. They will learn about parameters such as reactor volume, residence time, and catalyst loading, and how these affect conversion, selectivity, and efficiency. Real-world examples and case studies will be presented to provide students with practical insights into reactor performance evaluation.

By the end of this subchapter, students will have a solid foundation in reactor performance metrics. They will be equipped with the knowledge and tools needed to evaluate and optimize reactor performance, ultimately contributing to the advancement of chemical engineering and the efficient use of chemical reactions in various industries.

Residence Time Distribution

In the world of chemical engineering, understanding how reactants move and interact within a system is crucial. The concept of residence time distribution plays a vital role in this understanding. It allows us to analyze and optimize chemical reactions by examining the time it takes for reactants to leave a particular reactor or system.

Residence time refers to the average amount of time a molecule spends inside a reactor or system. It is a fundamental parameter that affects the performance and efficiency of chemical reactions. The residence time distribution (RTD) provides valuable information about how reactants are distributed in a system based on their residence times.

The RTD curve is a graphical representation that shows the distribution of residence times for different molecules in a reactor. It helps engineers assess the efficiency of reactors by revealing whether the reaction is well-mixed or exhibits preferential flow paths. By analyzing the shape of the curve, engineers can identify any potential issues or inefficiencies in the system.

Understanding the RTD is particularly important in the design and optimization of chemical reactors. Engineers need to ensure that reactants have sufficient time to react and that the desired products are formed. By manipulating the residence time, engineers can control reaction rates, conversion efficiencies, and product yields.

Several factors can influence the residence time distribution within a system. These include reactor design, flow rates, mixing, and the physical properties of the reactants. Different reactors, such as continuous flow reactors or batch reactors, will exhibit different residence time distributions. Therefore, it is essential for chemical engineering students to familiarize themselves with the various reactor types and their corresponding RTD characteristics.

Overall, residence time distribution is a fundamental concept in chemical engineering. It allows students to analyze and optimize chemical reactions by understanding the movement and interaction of reactants within a system. By manipulating the residence time, engineers can improve reaction performance and achieve desired product outcomes. Understanding the factors that influence residence time distribution is crucial for the successful design and operation of chemical reactors.

Catalysts and Catalysis

In the world of chemical engineering, understanding the role of catalysts and catalysis is crucial for students seeking to unleash the power of chemical reactions. Catalysts are substances that facilitate and accelerate chemical reactions without being consumed in the process. They play a fundamental role in numerous industrial processes, ranging from the production of fuels and pharmaceuticals to the creation of sustainable materials.

The concept of catalysis can be compared to a key that unlocks a reaction's potential. In the absence of a catalyst, reactions may occur, but at a slower rate and with higher energy requirements. Catalysts, on the other hand, lower the activation energy required for a reaction to take place. This means that reactions can occur under milder conditions, such as lower temperatures and pressures, resulting in significant energy savings and cost reductions.

Catalysts can be classified into two main categories: homogeneous and heterogeneous. Homogeneous catalysts are present in the same phase as the reactants, while heterogeneous catalysts exist in a different phase. Heterogeneous catalysts are more commonly used in industrial applications due to their ease of separation from the reaction mixture. They are often in the form of solid materials with high surface areas, such as metals, metal oxides, or zeolites.

The mechanism by which catalysts promote reactions is known as catalysis. There are three main types of catalysis: adsorption, surface reaction, and desorption. In the adsorption step, reactant molecules adhere to the catalyst's surface. This is followed by surface reactions, where the reactants undergo chemical transformations on the catalyst's surface. Finally, desorption occurs when the products detach from the catalyst.

Understanding catalysis is essential for designing efficient chemical processes. Students studying chemical engineering must comprehend the factors that influence catalytic activity, such as temperature, pressure, and the catalyst's physical and chemical properties. They must also be familiar with catalyst deactivation, which occurs when catalysts lose their activity over time due to poisoning, fouling, or sintering.

Moreover, the field of catalysis is continuously evolving, with ongoing research focusing on developing novel catalysts and optimizing existing ones. This research aims to enhance the selectivity and efficiency of reactions, reduce environmental impact, and promote sustainability in chemical engineering processes.

In summary, catalysts and catalysis are integral to the field of chemical engineering. They allow reactions to occur at faster rates and under milder conditions, leading to significant energy and cost savings. Understanding the different types of catalysts, their mechanisms, and the factors influencing their activity is crucial for students seeking to excel in chemical engineering and unleash the power of chemical reactions. By harnessing the potential of catalysis, students can contribute to the development of sustainable and efficient chemical processes, shaping the future of the industry.

Reaction Mechanisms and Rate Laws

Understanding the fundamental principles of reaction mechanisms and rate laws is crucial for students pursuing a career in chemical engineering. In this subchapter, we will delve into the intricate world of chemical reactions, exploring how they occur and the factors that influence their rates.

Chemical reactions are the backbone of the chemical engineering field, as they are responsible for the transformation of raw materials into valuable products. To comprehend these reactions, it is essential to grasp the concept of reaction mechanisms. A reaction mechanism describes the step-by-step process by which reactant molecules transform into products. By studying these mechanisms, chemical engineers can optimize reaction conditions to achieve desired yields and selectivity.

In addition to understanding reaction mechanisms, rate laws play a vital role in predicting the rate at which a reaction occurs. Rate laws express the relationship between the concentrations of reactants and the rate of reaction. By determining the rate law, engineers can design and control chemical processes more effectively. In this subchapter, we will explore the various types of rate laws and the mathematical expressions used to describe them.

Moreover, we will discuss the factors that influence reaction rates, such as temperature, pressure, concentration, and catalysts. Chemical engineers must have a deep understanding of these factors to design and optimize industrial-scale reactions. We will explore how changes in these parameters affect the rate of reaction and the overall efficiency of the process.

Throughout this subchapter, real-life examples and case studies will be provided to illustrate the practical applications of reaction mechanisms and rate laws in the field of chemical engineering. These examples will highlight the importance of understanding these concepts in diverse industries, such as pharmaceuticals, petrochemicals, and environmental engineering.

By the end of this subchapter, students will have a comprehensive understanding of the fundamental principles governing chemical reactions. They will be equipped with the knowledge and tools necessary to predict and optimize reaction rates, enabling them to tackle real-world engineering challenges with confidence.

Whether you are a budding chemical engineer or simply fascinated by the power of chemical reactions, this subchapter will provide you with the essential knowledge to unleash the potential of these transformative processes. So, let's embark on this exciting journey into the world of reaction mechanisms and rate laws – a journey that will unlock the power of chemical reactions and inspire the engineer within you.

Chapter 6: Transport Phenomena in Chemical Reactions

Mass Transfer

In the field of chemical engineering, the process of mass transfer plays a crucial role in various industrial applications. Mass transfer refers to the movement of substances from one phase to another, such as from a liquid to a gas or vice versa. This phenomenon is vital in many industrial processes, including distillation, absorption, extraction, and drying. Understanding the principles of mass transfer is therefore essential for students pursuing a career in chemical engineering.

One of the key concepts in mass transfer is diffusion. Diffusion is the spontaneous movement of molecules from an area of high concentration to an area of low concentration. This process occurs in both liquids and gases and is governed by Fick's law. Fick's law quantifies the rate of diffusion and provides a mathematical framework for understanding how concentration gradients influence mass transfer.

Another important aspect of mass transfer is interfacial mass transfer. Interfacial mass transfer occurs at the interface between two phases, such as a gas-liquid or liquid-liquid interface. It involves the transfer of mass across this interface, which can be influenced by factors such as surface area, temperature, and concentration gradients. Understanding interfacial mass transfer is crucial in designing efficient separation processes, such as distillation columns or liquid-liquid extraction units.

Students studying chemical engineering will also encounter mass transfer operations such as absorption and adsorption. Absorption refers to the process of transferring a solute from a gas phase into a liquid phase. This operation is commonly used in gas scrubbing to remove pollutants from industrial effluents. Adsorption, on the other hand, involves the attachment of molecules to a surface. This process is widely used in technologies like activated carbon filters to remove impurities from water or air.

In addition to diffusion, interfacial mass transfer, absorption, and adsorption, students will also learn about other mass transfer mechanisms like evaporation, crystallization, and membrane separation. Each of these mechanisms plays a vital role in different industrial applications, and understanding their underlying principles is crucial for chemical engineers.

In conclusion, mass transfer is a fundamental concept in chemical engineering. It encompasses various phenomena like diffusion, interfacial mass transfer, absorption, and adsorption, which are essential in numerous industrial processes. By studying mass transfer, students will gain the necessary knowledge and skills to design and optimize separation processes, contributing to the advancement of the chemical engineering field.

Heat Transfer

Heat transfer is a fundamental concept in the field of chemical engineering. It plays a crucial role in various industrial processes, including the design and operation of chemical reactors, heat exchangers, and distillation columns. Understanding the principles of heat transfer is essential for students pursuing a degree in chemical engineering, as it enables them to analyze and optimize heat transfer operations in various chemical processes.

In this subchapter, we will delve into the different modes of heat transfer, namely conduction, convection, and radiation. We will explore each mode in detail, discussing their mechanisms, mathematical formulations, and practical applications. Additionally, we will highlight the importance of heat transfer in chemical reactions and the role it plays in determining reaction rates and overall process efficiency.

Firstly, we will focus on conduction, which is the transfer of heat through a solid or stationary medium. We will explain the concept of thermal conductivity and discuss how it affects heat transfer rates. We will also introduce Fourier's law of heat conduction and provide examples of its application in various engineering systems.

Next, we will explore convection, which involves the transfer of heat through a fluid medium. We will discuss the different types of convection, namely natural convection and forced convection, and their respective heat transfer coefficients. We will also introduce the concept of boundary layers and their significance in convection heat transfer.

Lastly, we will discuss radiation, which is the transfer of heat through electromagnetic waves. We will explain the Stefan-Boltzmann law and the concept of emissivity, which determines the radiation properties of a material. We will also illustrate the application of radiation heat transfer in industrial processes, such as drying and thermal processing.

Throughout this subchapter, we will provide real-world examples and case studies, allowing students to connect theory with practical applications. We will also incorporate problem-solving exercises to reinforce understanding and encourage critical thinking.

By the end of this subchapter, students will have a comprehensive understanding of heat transfer principles and their applications in chemical engineering. They will be equipped with the knowledge and tools necessary to analyze and optimize heat transfer operations in various chemical processes, thereby unleashing the power of chemical reactions.

Momentum Transfer

In the field of Chemical Engineering, understanding the principles of momentum transfer is crucial for mastering the art of harnessing the power of chemical reactions. The concept of momentum transfer plays a central role in various engineering applications, ranging from fluid mechanics to heat transfer and mass transport. By comprehending the principles behind momentum transfer, students of Chemical Engineering can unlock a deeper understanding of how chemical reactions can be optimized and controlled.

At its core, momentum transfer refers to the movement of mass through a fluid or across a boundary. It encompasses the study of how momentum is transferred from one point to another, either within a fluid or between different phases of matter. This transfer of momentum is influenced by factors such as fluid viscosity, density, and velocity, as well as the properties of the object or boundary through which the fluid is flowing.

One fundamental aspect of momentum transfer is the study of fluid mechanics. Fluids, whether they are liquids or gases, display unique behaviors when it comes to momentum transfer. The study of fluid mechanics delves into the principles governing fluid flow, such as Bernoulli's equation, conservation of mass, and the Navier-Stokes equations. By understanding these principles, students can analyze and predict how fluids will behave in various chemical engineering processes, such as the design of pipelines, pumps, and mixing vessels.

Another vital application of momentum transfer is in heat transfer. Heat transfer is the study of how thermal energy is transferred between objects or regions with different temperatures. This process involves both the transfer of heat and the accompanying momentum transfer.

Understanding the principles of momentum transfer helps students comprehend how heat is transported through conduction, convection, and radiation. This knowledge is essential for designing and optimizing heat exchangers, distillation columns, and other heat transfer equipment, as well as for controlling chemical reactions that are strongly influenced by temperature changes.

Furthermore, the study of momentum transfer is crucial when it comes to mass transport. Mass transport involves the movement of chemical species or particles within a fluid or across a boundary. By understanding the principles of momentum transfer, students can predict and control the rate at which mass is transported within a system. This knowledge is vital for designing efficient separation processes, such as distillation, absorption, and extraction, as well as for optimizing reaction rates in chemical reactors.

In conclusion, momentum transfer is a fundamental concept in Chemical Engineering that underpins several critical processes and phenomena. By grasping the principles of momentum transfer, students can gain a deeper understanding of fluid mechanics, heat transfer, and mass transport. This knowledge empowers them to design and optimize chemical engineering processes effectively, unleashing the power of chemical reactions.

Chapter 7: Multiphase Reactors

Gas-Liquid Reactors

In the world of chemical engineering, gas-liquid reactors play a crucial role in the transformation of raw materials into valuable products. These reactors are designed to facilitate the interaction between gases and liquids, allowing for efficient and controlled chemical reactions to take place. Understanding the principles and applications of gas-liquid reactors is essential for students pursuing a career in chemical engineering.

One of the primary advantages of gas-liquid reactors is their ability to enhance mass transfer between gas and liquid phases. This is achieved through the introduction of gas bubbles into a liquid medium or by dispersing liquid droplets in a gas phase. By increasing the interfacial area between the gas and liquid, the reaction rate can be significantly improved. This makes gas-liquid reactors ideal for processes that involve gas absorption, stripping, or chemical reactions.

The design of gas-liquid reactors depends on various factors, including the desired reaction kinetics, reactant properties, and process conditions. Some common types of gas-liquid reactors include stirred tank reactors, bubble column reactors, and packed bed reactors. Each of these reactors has its own advantages and limitations, and the choice of reactor type depends on the specific needs of the process.

Stirred tank reactors, also known as agitated reactors, are widely used due to their versatility and ease of operation. They consist of a vessel equipped with impellers or stirrers that promote mixing and enhance mass transfer between the gas and liquid phases. Bubble column reactors, on the other hand, rely on the upward flow of gas bubbles through a liquid column. These reactors are particularly suitable for high gas-liquid mass transfer rates and are commonly used in wastewater treatment and gas absorption processes.

Packed bed reactors, as the name suggests, contain a bed of solid particles that are packed in a column. The gas and liquid phases flow through this bed, allowing for intimate contact and efficient mass transfer. Packed bed reactors are commonly used in catalytic reactions and are highly effective in terms of conversion and selectivity.

By understanding the principles and applications of gas-liquid reactors, students will gain insight into the design and operation of chemical processes that involve gas-liquid interactions. This knowledge will enable them to optimize reaction conditions, enhance reaction rates, and improve overall process efficiency. Gas-liquid reactors are essential tools in the field of chemical engineering, and mastering their concepts is crucial for students aspiring to become successful chemical engineers.

Gas-Solid Reactors

Gas-solid reactors are an integral part of the field of chemical engineering. Understanding the principles and applications of these reactors is crucial for students pursuing a career in this discipline. In this subchapter, we will delve into the world of gas-solid reactors and explore their significance in the realm of chemical reactions.

Gas-solid reactors are used in numerous industrial processes, including catalytic cracking, combustion, gasification, and fluidized bed reactors. These reactors involve the interaction between gases and solid particles, where chemical reactions take place on the surface of the solids. The primary purpose of gas-solid reactors is to facilitate the conversion of reactants into desired products efficiently and economically.

One of the key concepts in gas-solid reactors is the contact between the gas and solid phases. The efficiency of this contact greatly influences the reaction rate and overall reactor performance. Various factors such as particle size, shape, and distribution, as well as gas flow rate and temperature, need to be considered to optimize the contact between the two phases. Students will learn about the different modes of gas-solid contact, including fixed bed, fluidized bed, and moving bed reactors, and the advantages and limitations of each.

Furthermore, this subchapter will delve into the design and operation principles of gas-solid reactors. Students will gain insights into reactor sizing, heat and mass transfer mechanisms, and reactor performance evaluation. The importance of catalysts in gas-solid reactions will also be emphasized, as they play a crucial role in enhancing the reaction rate and selectivity.

Case studies and real-life examples will be incorporated throughout the subchapter to provide students with a practical understanding of gas-solid reactors. These examples will showcase the application of gas-solid reactors in industries such as petroleum refining, chemical production, and environmental engineering.

By the end of this subchapter, students will have a comprehensive understanding of gas-solid reactors, their design principles, and their applications in chemical engineering. They will be equipped with the knowledge and skills necessary to analyze, design, and optimize gas-solid reactors for various industrial processes. This subchapter will not only enhance their theoretical understanding of chemical reactions but also prepare them for the challenges they may encounter in their future careers as chemical engineers.

Liquid-Liquid Reactors

In the fascinating world of chemical engineering, the study of chemical reactions forms the cornerstone of understanding and manipulating various substances to create new products and enhance existing ones. One intriguing area of research in this field is liquid-liquid reactors. These reactors are specifically designed to handle reactions occurring between two immiscible liquids, offering unique advantages and challenges.

Liquid-liquid reactions involve the interaction between two liquid phases that do not mix together, such as oil and water. These reactions often occur in industries such as pharmaceuticals, food processing, and petroleum, where the desired reaction takes place at the interface between the two liquids. The primary goal of liquid-liquid reactors is to maximize the interfacial area, allowing for efficient mass transfer and reaction rates.

One common type of liquid-liquid reactor is the continuous stirred tank reactor (CSTR), which involves continuously mixing the two immiscible liquids in a tank. This reactor design allows for thorough mixing and ensures a large interfacial area, promoting faster reactions. Another type is the liquid-liquid extraction column, which utilizes the countercurrent flow of the two immiscible liquids to enhance contact and reaction efficiency.

To optimize the performance of liquid-liquid reactors, engineers must consider several factors. The choice of solvent, reactant concentrations, temperature, and agitation rate significantly influence the reaction kinetics and the separation of products. Additionally, the choice of reactor design, such as the type and size, plays a crucial role in achieving desired outcomes.

Liquid-liquid reactions present unique challenges compared to reactions involving homogeneous or heterogeneous systems. The separation of the desired product from the immiscible liquids can be complex and requires careful design and consideration. Engineers must also address issues such as emulsion formation, phase inversion, and coalescence to ensure efficient and economical operations.

Studying liquid-liquid reactors is of utmost importance for students pursuing chemical engineering. Understanding the principles, design considerations, and challenges associated with these reactors will equip them with the knowledge and skills necessary to overcome practical problems in various industries. Moreover, this knowledge will enable students to explore innovative approaches and develop novel solutions to enhance reaction efficiency and product quality.

In conclusion, liquid-liquid reactors provide a fascinating area of study within chemical engineering. The ability to manipulate reactions occurring between immiscible liquids opens up opportunities for numerous industries. By delving into the principles, design considerations, and challenges of liquid-liquid reactors, students can unleash the power of chemical reactions and contribute to the advancement of various fields, making them valuable assets in the chemical engineering community.

Chapter 8: Safety in Chemical Reactions

Hazards and Risk Assessment

In the field of chemical engineering, it is essential to understand and manage the potential hazards associated with the handling and processing of chemicals. Hazards can arise from a wide range of sources, including toxic substances, flammable materials, reactive chemicals, and high-pressure systems. Therefore, a thorough understanding of these hazards and proper risk assessment techniques are crucial to ensure the safety of both personnel and the environment.

This subchapter will delve into the various hazards encountered in chemical engineering and provide an overview of risk assessment methods utilized by professionals in the field. By addressing these topics, students pursuing chemical engineering will gain a comprehensive understanding of the potential dangers they may encounter in their future careers.

The subchapter will begin by exploring the different types of hazards commonly found in chemical engineering processes. This will include discussions on the toxicological properties of chemicals, potential fire and explosion risks, as well as the reactivity of certain substances. Real-life examples and case studies will be included to illustrate the importance of hazard identification and mitigation.

Next, the subchapter will shift its focus to risk assessment techniques. Students will learn about quantitative and qualitative risk assessment methods and how these approaches help identify and prioritize potential hazards. The content will also emphasize the significance of risk management strategies and the use of protective systems, such as containment, ventilation, and personal protective equipment.

Furthermore, the subchapter will highlight the role of process safety management in chemical engineering. Students will be introduced to concepts like hazard analysis, process safety audits, and emergency response planning. These aspects are critical for ensuring that chemical processes are designed, operated, and maintained in a manner that minimizes risks to personnel, the community, and the environment.

To consolidate the knowledge gained, practical exercises and self-assessment questions will be included throughout the subchapter. These interactive elements will encourage students to apply hazard and risk assessment techniques to real-world scenarios, enhancing their understanding and problem-solving skills.

By the end of this subchapter, students will have a solid foundation in hazard identification, risk assessment, and process safety management. Armed with this knowledge, they will be well-equipped to navigate the challenges of chemical engineering and contribute to the development of safe and sustainable chemical processes.

Safety Measures and Protocols

Chemical Engineering for Students: Unleashing the Power of Chemical Reactions

As students aspiring to become future chemical engineers, it is essential to understand the importance of safety measures and protocols in the field of chemical engineering. Working with various chemicals and conducting experiments involving chemical reactions requires meticulous attention to safety to prevent accidents and ensure the well-being of both individuals and the environment. In this subchapter, we will delve into the fundamental safety measures and protocols that every chemical engineering student should be aware of.

One of the primary safety measures is wearing appropriate personal protective equipment (PPE). This includes lab coats, safety goggles, gloves, and closed-toe shoes, which act as a barrier between chemicals and the body. PPE helps protect against chemical splashes, spills, and exposure, reducing the risk of injuries or chemical burns.

Furthermore, understanding the properties and hazards associated with different chemicals is crucial. Chemicals can vary in toxicity, reactivity, flammability, and explosiveness. Students must be well-versed in reading safety data sheets (SDS) to identify potential risks and follow recommended precautions for handling each chemical.

Proper handling and storage of chemicals are equally important. Chemicals should be stored in designated areas, away from incompatible substances, heat sources, or direct sunlight. Additionally, each chemical container should be appropriately labeled, indicating its contents, hazards, and safety precautions.

When conducting experiments, it is essential to work in a well-ventilated area or employ fume hoods to prevent the inhalation of toxic fumes. Chemical reactions should be carried out on stable surfaces, and students should be cautious when handling glassware or sharp objects to avoid accidents.

Emergency preparedness is another vital aspect of safety protocols. Students should be aware of the location of emergency exits, fire extinguishers, and safety showers in their laboratory. Additionally, they should have a clear understanding of evacuation procedures and know how to respond to different emergency scenarios, such as chemical spills or fires.

Regular equipment maintenance and calibration are essential for ensuring accurate readings and preventing equipment malfunctions that could lead to accidents. Students should be diligent in reporting any faulty equipment to their professors or lab supervisors.

By adhering to these safety measures and protocols, chemical engineering students can create a safe and conducive environment for themselves and those around them. Remember, safety should always be the utmost priority to prevent accidents, protect lives, and unleash the power of chemical reactions responsibly.

Emergency Response and Incident Management

In the fast-paced world of chemical engineering, it is crucial for students to be equipped with the knowledge and skills necessary to handle emergency situations effectively. The subchapter on Emergency Response and Incident Management in our book, "Chemical Engineering for Students: Unleashing the Power of Chemical Reactions," provides valuable insights into this critical aspect of the field.

Understanding the fundamentals of emergency response is of utmost importance for students pursuing a career in chemical engineering. This subchapter delves into the various types of emergencies that can occur in a chemical engineering setting, including fires, chemical spills, and equipment failures. By studying these scenarios, students can develop a proactive mindset and anticipate potential risks, enabling them to respond swiftly and decisively in times of crisis.

One of the key areas covered in this subchapter is risk assessment and management. Students will learn how to identify potential hazards, evaluate their consequences, and implement appropriate control measures to mitigate risks. By examining case studies and real-life examples, students will gain a comprehensive understanding of the importance of risk assessment and management in chemical engineering.

Moreover, this subchapter also emphasizes the significance of effective communication during emergencies. Students will learn how to establish clear communication channels, both within the organization and with external stakeholders, such as emergency responders and regulatory bodies. They will also explore the importance of developing emergency response plans and conducting regular drills to ensure a coordinated and efficient response in critical situations.

Furthermore, this subchapter highlights the role of incident management in chemical engineering. Students will understand the importance of establishing incident command systems, which involve assigning roles and responsibilities to key personnel, establishing incident priorities, and implementing strategies to mitigate the impact of an incident. They will also explore the ethical and legal considerations associated with incident management, including the importance of prioritizing human safety and minimizing environmental impact.

In summary, the subchapter on Emergency Response and Incident Management in "Chemical Engineering for Students: Unleashing the Power of Chemical Reactions" equips students with the knowledge and skills necessary to handle emergencies in the field of chemical engineering. By studying risk assessment, effective communication, and incident management, students can develop a proactive mindset and contribute to the safe and sustainable practices of the industry.

Chapter 9: Industrial Applications of Chemical Reactions

Petrochemical Industry

The Petrochemical Industry: A Catalyst for Chemical Engineering Students

As chemical engineering students, it is crucial to understand the role and significance of the petrochemical industry in today's world. The petrochemical industry plays a vital role in the production of a wide range of chemicals and materials that are essential for our daily lives. From plastics to synthetic fibers, pharmaceuticals to fertilizers, the petrochemical industry serves as the backbone of the modern chemical engineering field.

The petrochemical industry primarily relies on the transformation of crude oil and natural gas into various chemical compounds through a series of sophisticated processes. These processes involve converting hydrocarbons obtained from crude oil and natural gas into valuable products using chemical reactions. Chemical engineers are at the forefront of designing and optimizing these processes to ensure maximum efficiency and sustainability.

One of the key processes in the petrochemical industry is cracking, which involves breaking down larger hydrocarbon molecules into smaller ones. This process is essential for the production of various chemicals, such as ethylene and propylene, which serve as building blocks for the manufacturing of plastics, synthetic rubbers, and other materials. Chemical engineers play a crucial role in developing and improving cracking technologies, enabling the industry to produce a diverse range of products.

Another vital aspect of the petrochemical industry is polymerization, where small molecules, such as ethylene, are chemically bonded together to form long chains called polymers. Polymerization is responsible for the production of various types of plastics, including polyethylene and polypropylene, which have revolutionized industries such as packaging, construction, and automotive. Chemical engineers are involved in optimizing polymerization processes to enhance the properties and performance of these materials.

Moreover, the petrochemical industry also encompasses the production of various specialty chemicals and intermediates used in the pharmaceutical, agricultural, and personal care industries. Chemical engineers in this field work on developing innovative processes for the synthesis and purification of these specialty chemicals, ensuring their quality and purity.

In addition to the technical aspects, it is crucial for chemical engineering students to understand the environmental and sustainability challenges associated with the petrochemical industry. As the industry heavily relies on fossil fuels, it is essential to explore alternative feedstocks and develop greener processes. Chemical engineers are actively involved in researching and implementing sustainable solutions, such as bio-based feedstocks, renewable energy sources, and carbon capture technologies.

In conclusion, the petrochemical industry serves as a catalyst for chemical engineering students, offering a vast arena for innovation and growth. Understanding the intricacies of this industry is essential for aspiring chemical engineers, as it provides them with the knowledge and skills to contribute to the development of sustainable and efficient processes. By unleashing the power of chemical reactions, students can play a vital role in shaping the future of the petrochemical industry, making it more environmentally friendly and economically viable.

Pharmaceutical Industry

The pharmaceutical industry plays a vital role in improving human health and well-being. It combines the principles of chemistry, biology, and engineering to develop and produce life-saving medications. In this subchapter, we will explore the fascinating world of the pharmaceutical industry and delve into the key aspects that make it a unique field within chemical engineering.

One of the primary goals of the pharmaceutical industry is to discover and develop new drugs. Chemical engineers in this industry work tirelessly to identify potential drug candidates, design efficient synthesis routes, and optimize manufacturing processes. They utilize their knowledge of chemical reactions, thermodynamics, and transport phenomena to create safe and effective medications.

The process of drug discovery and development involves several stages, starting from the initial identification of a target disease, followed by extensive research and testing. Chemical engineers play a crucial role in formulating and testing drug candidates, ensuring their stability, efficacy, and safety. They design experiments, analyze data, and collaborate with other scientists to advance the drug development process.

Once a drug candidate is identified, it undergoes rigorous clinical trials to assess its effectiveness and potential side effects. Chemical engineers contribute to these trials by analyzing the pharmacokinetics and pharmacodynamics of the drug, ensuring that it reaches the desired target in the body and produces the desired therapeutic effect.

Moreover, chemical engineers in the pharmaceutical industry are responsible for scaling up the production of drugs from laboratory-scale to commercial-scale. They optimize the manufacturing process to ensure efficiency, cost-effectiveness, and safety. This involves designing reactors, separation processes, and purification techniques to produce high-quality pharmaceutical products.

The pharmaceutical industry also faces unique challenges, such as patent protection, regulatory compliance, and quality control. Chemical engineers need to ensure that the manufacturing process adheres to strict regulations and that the final product meets the required standards. They work closely with regulatory authorities to ensure the safety and efficacy of medications.

In summary, the pharmaceutical industry offers immense opportunities for chemical engineers to contribute to the development and production of life-saving drugs. This subchapter will provide a comprehensive overview of the industry, from drug discovery to manufacturing and regulatory compliance. It will equip students with a deep understanding of the pharmaceutical field and inspire them to pursue careers in this vital industry.

Food and Beverage Industry

The food and beverage industry is a vital sector that plays a significant role in our daily lives. As students of chemical engineering, it is essential to understand the intricacies and challenges faced by this industry. This subchapter aims to provide an overview of the food and beverage industry and shed light on the various chemical engineering applications within this field.

The food and beverage industry encompasses a wide range of processes, including food processing, manufacturing, packaging, and distribution. It involves the transformation of raw materials, such as fruits, vegetables, grains, and meat, into safe and nutritious products for consumption. Chemical engineering principles are fundamental in optimizing these processes, ensuring product quality, and meeting regulatory standards.

One of the key areas where chemical engineering plays a pivotal role is in food processing. This involves the application of physical and chemical transformations to raw materials to create value-added products. Chemical engineers work on developing efficient and sustainable processes, such as extraction, fermentation, and thermal processing, to maximize yield, minimize waste, and improve the overall quality of food products.

Another crucial aspect of the food and beverage industry is packaging. Chemical engineers contribute to the design and development of packaging materials that preserve product freshness, prevent contamination, and extend shelf life. They also focus on ensuring the safety of packaging materials, considering factors such as migration of chemicals and the impact on the environment.

Additionally, chemical engineers are involved in the development and optimization of food additives and flavorings. These additives enhance the taste, texture, and appearance of food products, providing consumers with a pleasant sensory experience. Chemical engineers play a vital role in researching and developing these additives, as well as ensuring their safety and regulatory compliance.

Moreover, sustainability and waste management are pressing concerns within the food and beverage industry. Chemical engineers work towards minimizing waste generation, optimizing energy consumption, and implementing environmentally friendly practices. They explore innovative technologies, such as bioremediation, waste valorization, and water treatment, to mitigate the industry's ecological footprint.

In conclusion, the food and beverage industry offers a plethora of opportunities for chemical engineers to make a significant impact. From food processing to packaging and sustainability, chemical engineering principles are essential in optimizing processes, ensuring product quality, and meeting regulatory standards. As students, understanding the nuances of this industry equips us with the knowledge and skills to contribute meaningfully to its growth and development.

Environmental Engineering

Environmental engineering is a crucial field within the realm of chemical engineering that focuses on the preservation, protection, and improvement of the environment. It plays a significant role in tackling the challenges posed by pollution, waste management, and sustainable development. In this subchapter, we will explore the fundamental principles and applications of environmental engineering, providing students with the knowledge and skills needed to address pressing environmental concerns.

One of the key areas of environmental engineering is the study and management of air pollution. Students will delve into the various sources of air pollutants, such as industrial emissions, transportation, and energy production. They will learn about the harmful effects of air pollution on human health and the environment, as well as the different techniques and technologies used to control and reduce air pollution. This includes the design and operation of air pollution control systems and the implementation of regulations and policies to ensure clean air quality.

Another significant aspect of environmental engineering is water and wastewater treatment. Students will gain an understanding of the challenges associated with water scarcity, water pollution, and the need for sustainable water management. They will explore the principles and processes involved in water treatment, including filtration, disinfection, and desalination. Additionally, students will learn about wastewater treatment methods to remove contaminants and pollutants from domestic, industrial, and agricultural sources, ensuring the protection of water bodies and public health.

Students will also be introduced to waste management and the concept of the circular economy. They will learn about the different types of waste, including solid waste, hazardous waste, and electronic waste. The subchapter will cover waste reduction strategies, recycling techniques, and the safe disposal of waste. Furthermore, students will gain insight into the significance of resource recovery, emphasizing the importance of reusing and recycling materials to minimize environmental impact.

Lastly, sustainable development will be a focal point of this subchapter. Students will explore the principles of sustainability and its integration into engineering practices. They will learn about life cycle assessment, eco-design, and green chemistry, which contribute to the development of environmentally friendly processes and products. The subchapter will also highlight the role of environmental engineers in promoting renewable energy sources, energy efficiency, and overall environmental stewardship.

By the end of this subchapter, students will have a comprehensive understanding of environmental engineering and its vital role in addressing global environmental challenges. They will be equipped with the knowledge and skills necessary to apply sustainable and environmentally conscious practices in their future chemical engineering careers, making a positive impact and contributing to a more sustainable world.

Chapter 10: Future Trends in Chemical Engineering

Advances in Reaction Engineering

Chemical reactions form the backbone of numerous industrial processes and are vital to the production of a wide range of products that we use in our daily lives. As students of chemical engineering, it is crucial to understand the advances in reaction engineering that have revolutionized the field and continue to shape the future of industry.

One significant advance in reaction engineering is the development of novel catalysts. Catalysts are substances that facilitate chemical reactions by lowering the activation energy required for the reaction to occur. Over the years, researchers have made remarkable progress in designing and optimizing catalysts to improve reaction rates and selectivity. From heterogeneous catalysts used in industrial processes to enzymatic catalysts employed in biotechnology, these advancements have opened new doors to more efficient and sustainable reactions.

Another exciting development is the integration of reaction engineering with computational modeling. The advent of powerful computers and sophisticated software has allowed engineers to simulate and predict reaction behavior with incredible accuracy. These simulations provide valuable insights into reaction mechanisms, reaction kinetics, and the optimization of reaction conditions. By leveraging computational tools, students can now explore and design chemical reactions in silico, saving time, resources, and reducing the need for extensive experimentation.

Furthermore, the field of reaction engineering has embraced the concept of green chemistry. This approach focuses on developing environmentally friendly processes that minimize waste generation, use renewable resources, and reduce energy consumption. Students can learn about techniques such as solvent replacement, process intensification, and the use of alternative reaction pathways to achieve greener and more sustainable chemical reactions. The integration of green chemistry principles in reaction engineering not only benefits the environment but also improves the efficiency and economics of industrial processes.

In recent years, there has been a growing interest in the field of bioreaction engineering. Bioreactions utilize living organisms, such as bacteria or enzymes, to catalyze chemical reactions. This emerging field has witnessed significant advancements in the development of bioreactors, genetic engineering techniques, and enzyme immobilization methods. Bioreactions offer numerous advantages, including high selectivity, mild reaction conditions, and the use of renewable feedstocks. Students exploring this niche can delve into topics like metabolic engineering, cell growth kinetics, and bioprocess optimization.

In conclusion, advances in reaction engineering have transformed the way chemical processes are designed and operated. From the development of innovative catalysts to the integration of computational modeling, green chemistry, and bioreaction engineering, these advancements have paved the way for more sustainable, efficient, and economically viable chemical reactions. As students of chemical engineering, understanding these advances is crucial for harnessing the power of chemical reactions and shaping a future that is sustainable and environmentally conscious.

Sustainable and Green Chemistry

In recent years, there has been a growing awareness of the impact that chemical processes have on the environment. As students studying chemical engineering, it is crucial for us to understand the principles of sustainable and green chemistry. This subchapter aims to explore these concepts and highlight their importance in our field.

Sustainable chemistry, also known as green chemistry, involves the design and development of chemical processes that minimize the use and generation of hazardous substances. It focuses on developing innovative solutions that not only meet our present needs but also ensure the well-being of future generations. By incorporating sustainable practices, we can reduce the environmental impact of chemical reactions while still achieving desired outcomes.

One fundamental principle of sustainable chemistry is the use of renewable resources. As students of chemical engineering, we must explore ways to replace non-renewable resources with sustainable alternatives. This can involve utilizing biomass, such as agricultural waste, as a feedstock for chemical reactions, or harnessing solar energy to power industrial processes. By doing so, we can reduce our reliance on fossil fuels and contribute to a more sustainable future.

Another aspect of sustainable chemistry is waste reduction. Traditional chemical processes often generate large amounts of hazardous byproducts. Through the application of green chemistry principles, we can minimize waste production by designing reactions that produce fewer unwanted byproducts. Additionally, we can explore ways to recycle and reuse materials, further reducing our environmental footprint.

Furthermore, sustainable chemistry emphasizes the use of safer chemicals. By selecting and designing chemicals that are less toxic or harmful to humans and the environment, we can mitigate potential risks associated with chemical processes. This includes the identification and development of alternative solvents, catalysts, and reaction conditions that are more environmentally friendly.

It is important for us, as students of chemical engineering, to be aware of the opportunities and challenges presented by sustainable and green chemistry. By incorporating these principles into our future careers, we can contribute to a more sustainable and environmentally conscious industry. This subchapter will delve deeper into the various strategies and techniques used in sustainable chemistry, highlighting case studies and real-world applications that demonstrate the positive impact of green chemical engineering.

In conclusion, sustainable and green chemistry is a vital aspect of chemical engineering. By adopting these principles, we can minimize the environmental impact of chemical processes, reduce waste generation, and develop safer alternatives. As students, it is our responsibility to embrace these concepts and apply them in our future careers, ensuring a more sustainable and greener future for our industry and the world.

Nanotechnology in Chemical Reactions

In recent years, the field of nanotechnology has made significant strides in revolutionizing various industries, and chemical engineering is no exception. Nanotechnology involves the manipulation and control of matter at the nanoscale, which is roughly the size of atoms and molecules. This emerging technology has proven to be a game-changer in chemical reactions, offering unprecedented control and efficiency in the design and synthesis of new materials.

One of the key applications of nanotechnology in chemical reactions is catalyst design. Catalysts are substances that speed up chemical reactions without being consumed in the process. Traditional catalysts often have limitations, such as low selectivity, high energy consumption, and limited reusability. However, with the advent of nanotechnology, engineers can now design catalysts with enhanced properties by precisely controlling their size, shape, and composition at the nanoscale.

Nanocatalysts have shown remarkable improvements in various chemical reactions. For example, in the field of environmental engineering, nanocatalysts have been developed to remove pollutants from air and water more efficiently. They can selectively break down harmful chemicals, such as volatile organic compounds and heavy metals, into harmless byproducts. This has significant implications for sustainable development and the preservation of our environment.

Furthermore, nanotechnology has enabled the development of smart materials and nanosensors that can revolutionize the monitoring and control of chemical reactions. These materials can respond to changes in temperature, pressure, pH, or even the presence of specific molecules, allowing for real-time monitoring and precise control over reaction conditions. This level of control not only improves the overall efficiency of chemical reactions but also reduces waste and enhances product quality.

Additionally, nanotechnology has opened up new possibilities in the field of drug delivery systems. By engineering nanoparticles with specific properties, such as size, shape, and surface chemistry, pharmaceutical engineers can improve the targeted delivery of drugs to specific cells or tissues, leading to enhanced therapeutic effects and reduced side effects.

In conclusion, nanotechnology has emerged as a powerful tool in chemical engineering, offering unprecedented control and efficiency in chemical reactions. From catalyst design to smart materials and drug delivery systems, nanotechnology has the potential to revolutionize the field and drive innovations that benefit society as a whole. As students of chemical engineering, understanding and harnessing the power of nanotechnology in chemical reactions will undoubtedly be crucial in shaping the future of this discipline.

Chapter 11: Case Studies and Projects

Case Study 1: Designing a Chemical Reactor for a Pharmaceutical Process

Welcome to the first case study in our book, "Chemical Engineering for Students: Unleashing the Power of Chemical Reactions." In this chapter, we will delve into the fascinating world of chemical engineering, specifically focusing on the niche of designing chemical reactors for pharmaceutical processes.

For students studying chemical engineering, understanding the intricacies of reactor design is essential. Chemical reactors are at the heart of any chemical process, and designing them to be efficient, safe, and cost-effective is paramount. In this case study, we will explore the challenges faced by chemical engineers in designing a chemical reactor for a pharmaceutical process.

Pharmaceutical processes require precise control and high purity to ensure the production of safe and effective drugs. Chemical engineers play a crucial role in developing reactors that can handle the complex chemistry involved in pharmaceutical synthesis. From selecting the right catalysts and reactants to optimizing reaction conditions, chemical engineers must carefully consider various factors to achieve the desired outcomes.

In this case study, we will follow the journey of a team of chemical engineers as they design a reactor for a pharmaceutical process. We will explore the selection of suitable reactor types, such as batch, continuous, or semi-batch, based on the specific requirements of the process. The team will also discuss the importance of heat transfer, mass transfer, and reaction kinetics in reactor design.

Furthermore, safety considerations are paramount in the pharmaceutical industry. We will explore how the team incorporates safety features such as pressure relief systems, temperature control mechanisms, and emergency shutdown procedures into their reactor design. Additionally, we will highlight the importance of process control and automation in ensuring the efficient operation of the reactor.

Throughout this case study, we will provide real-world examples and practical insights to help students understand the principles and challenges of designing a chemical reactor for a pharmaceutical process. By the end, readers will gain a comprehensive understanding of the thought process and decision-making involved in this niche of chemical engineering.

We hope that this case study will inspire and empower students to explore the exciting field of chemical engineering further. Stay tuned for more intriguing case studies and in-depth discussions as we continue to unleash the power of chemical reactions in our book, "Chemical Engineering for Students."

Case Study 2: Optimization of a Catalytic Reaction for Energy Production

Introduction:

In today's world, where energy demands are rapidly increasing, the optimization of chemical reactions plays a crucial role in meeting these requirements sustainably. This case study focuses on the optimization of a catalytic reaction for energy production, highlighting the importance of chemical engineering in the field of renewable energy.

Background:

Chemical engineering combines various scientific principles, engineering techniques, and chemical reactions to design and optimize processes that convert raw materials into valuable products. One such process is the catalytic reaction, which involves the use of catalysts to enhance the rate and selectivity of chemical reactions. Catalysts are substances that facilitate reactions without being consumed, thus making them ideal for sustainable energy production.

Objective:

The objective of this case study is to optimize a catalytic reaction for energy production, specifically targeting the conversion of renewable feedstocks into high-energy fuels. By manipulating reaction conditions, catalyst properties, and reactor design, chemical engineers can maximize the efficiency and yield of the desired product while minimizing unwanted by-products and energy consumption.

Methodology:

1. Selection of Feedstock: The first step is to identify a renewable feedstock, such as biomass or bioethanol, that can be converted into high-energy fuels through the catalytic reaction.

2. Catalyst Design: Chemical engineers will then design and synthesize a catalyst that exhibits high activity, selectivity, and stability for the desired reaction. This may involve modifying the catalyst's composition, structure, or surface properties.

3. Reaction Optimization: The reaction conditions, such as temperature, pressure, and residence time, will be optimized to maximize the conversion of the feedstock into the desired fuel while minimizing side reactions and catalyst deactivation.

4. Reactor Design: Chemical engineers will select an appropriate reactor type, such as a fixed-bed, fluidized-bed, or continuous-flow reactor, considering factors like mass transfer limitations, heat transfer requirements, and catalyst regeneration.

Results   and   Benefits:
Through optimization, the catalytic reaction for energy production can achieve higher fuel yields, reduced energy consumption, and minimized environmental impact. This not only contributes to the development of sustainable energy sources but also offers economic benefits through the production of clean and efficient fuels.

Conclusion:
The optimization of catalytic reactions for energy production is a key aspect of chemical engineering in the field of renewable energy. By leveraging their knowledge of reaction kinetics, catalyst design, and process optimization, chemical engineers play a vital role in developing sustainable energy solutions that meet the increasing global energy demands. This case study highlights the importance of chemical engineering principles in harnessing the power of chemical reactions for a greener future.

Project: Designing an Eco-Friendly Process for Chemical Synthesis

Chapter 4: Project: Designing an Eco-Friendly Process for Chemical Synthesis

Introduction:

Welcome to the exciting world of chemical engineering! In this chapter, we will embark on a thrilling project aimed at designing an eco-friendly process for chemical synthesis. As future chemical engineers, it is crucial to learn how to utilize the power of chemical reactions while minimizing their impact on the environment. This project will challenge your knowledge and skills, allowing you to apply sustainable principles to the field of chemical engineering.

Understanding Eco-Friendly Chemical Synthesis: Chemical synthesis is the backbone of the chemical industry, but traditional methods often generate large amounts of waste and contribute to pollution. Our task is to design a process that minimizes the use of hazardous materials, reduces energy consumption, and maximizes the yield of desired products. By doing so, we can contribute to a greener and more sustainable future.

Step 1: Research and Analysis: Begin by researching different chemical synthesis techniques and identifying the environmental challenges associated with them. Look for alternative reactions, catalysts, and solvents that can reduce waste generation and improve efficiency. Analyze the life cycle of each process, considering factors such as raw material extraction, transportation, and disposal.

Step 2: Conceptual Design: Based on your research, develop a conceptual design for an eco-friendly chemical synthesis process. Consider factors such as reaction conditions, catalyst selection, solvent choice, and process optimization. Use simulation software to model the proposed process and analyze its environmental impact, energy consumption, and product yield.

Step 3: Process Optimization: Refine your design by optimizing key parameters such as temperature, pressure, and reactant concentrations. Explore the use of renewable feedstocks, green solvents, and energy-efficient technologies. Incorporate process safety measures to ensure a sustainable and safe operation.

Step 4: Economic Analysis: Evaluate the economic feasibility of your eco-friendly process. Compare it with traditional methods, considering factors such as capital costs, operating expenses, and market potential. Justify the advantages of your design in terms of cost savings, environmental benefits, and potential market demand.

Conclusion:
By undertaking this project, you will gain valuable insights into the challenges and opportunities of designing eco-friendly processes in chemical engineering. This project not only enhances your technical skills but also instills a sense of responsibility towards the environment. As students of chemical engineering, it is our duty to unleash the power of chemical reactions while preserving the planet for future generations. Together, let's revolutionize the chemical industry and pave the way for a greener and more sustainable future.

Chapter 12: Conclusion and Career Opportunities

Summary of Key Concepts

In this subchapter, we will provide a concise summary of the key concepts covered in the book "Chemical Engineering for Students: Unleashing the Power of Chemical Reactions." Designed specifically for students in the field of chemical engineering, this summary serves as a valuable resource to reinforce your understanding of the fundamental principles and applications of chemical reactions.

1. Chemical Reactions: Understanding the basics Chemical reactions form the foundation of chemical engineering. This section introduces the concept of chemical reactions, including the role of reactants, products, and the stoichiometry involved. You will learn about balancing chemical equations, identifying reaction types, and the fundamental laws governing chemical transformations.

2. Thermodynamics: The driving force Thermodynamics plays a crucial role in chemical engineering. This section explores the laws of thermodynamics and their application to chemical reactions. You will gain an understanding of energy, enthalpy, entropy, and Gibbs free energy, which are essential for predicting and optimizing chemical reactions.

3. Reaction Kinetics: The rate of transformation Reaction kinetics focuses on the study of reaction rates and mechanisms. This section delves into the factors influencing reaction rates, including temperature, concentration, and catalysts. You will learn about rate laws, reaction order, and how to determine the rate constant. Furthermore, the concept of reaction mechanisms and their role in explaining complex reactions will be explored.

4.   Reactor   Design: Practical   applications Reactors are essential in chemical engineering for carrying out chemical reactions on an industrial scale. This section covers the design and operation of different types of reactors, including batch, continuous, and semi-batch reactors. You will understand the importance of reactor sizing, conversion, and selectivity in achieving desired reaction outcomes.

5.   Transport   Phenomena: Mass   and   heat   transfer Transport phenomena involve the movement of mass, energy, and momentum. This section introduces the principles of mass and heat transfer and their applications in chemical engineering. You will learn about diffusion, convection, and conduction, as well as how to calculate mass transfer coefficients and heat transfer rates.

6. Safety and Environmental Considerations: Responsible engineering Chemical engineers have a crucial role in ensuring safe and sustainable practices. This section addresses the importance of safety protocols, hazard identification, and risk assessment in chemical engineering processes. It also emphasizes the need for environmentally conscious approaches to minimize the impact of chemical reactions on the environment.

By understanding these key concepts, you will be equipped with a solid foundation in chemical engineering principles and their practical applications. Whether you are a student embarking on your academic journey or a professional looking to refresh your knowledge, "Chemical Engineering for Students: Unleashing the Power of Chemical Reactions" offers a comprehensive and accessible guide to the fascinating world of chemical engineering.

Career Paths in Chemical Engineering

Chemical Engineering is a diverse and dynamic field that offers a wide range of career opportunities. Whether you are interested in research and development, process design, or environmental sustainability, the field of chemical engineering has something for everyone. In this subchapter, we will explore some of the exciting career paths available to students in the field of chemical engineering.

One of the most common career paths for chemical engineering graduates is in the field of process engineering. Process engineers are responsible for designing and optimizing the production processes used in various industries such as pharmaceuticals, oil and gas, and food processing. They work closely with scientists, technicians, and operators to ensure that the production processes are efficient, safe, and environmentally friendly.

Another popular career choice for chemical engineering graduates is in research and development. In this role, engineers work on developing new products, processes, and technologies. They conduct experiments, analyze data, and collaborate with scientists and other engineers to innovate and improve existing products or develop new ones. Research and development careers are particularly exciting as they often involve working on cutting-edge technologies and finding solutions to complex problems.

Chemical engineers also play a crucial role in environmental sustainability. With growing concerns about climate change and pollution, there is an increasing demand for professionals who can develop sustainable processes and technologies. Environmental engineers focus on developing strategies to reduce waste, minimize energy consumption, and improve the overall environmental impact of industrial processes.

For those interested in the healthcare industry, careers in pharmaceuticals and biotechnology offer exciting opportunities. Chemical engineers in this field work on developing and optimizing processes for the production of pharmaceutical drugs, vaccines, and other medical products. They also contribute to the development of novel drug delivery systems, such as nanoparticles and microparticles, which enhance drug efficacy and patient outcomes.

In addition to these traditional career paths, chemical engineering graduates can also venture into entrepreneurship and start their own companies. With a strong background in problem-solving and technical expertise, chemical engineers are well-equipped to identify market gaps and develop innovative solutions to address them.

Overall, the field of chemical engineering offers a plethora of career opportunities for students. Whether you are passionate about research and development, process design, environmental sustainability, or entrepreneurship, a degree in chemical engineering can open doors to a rewarding and fulfilling career. So, if you are a student considering a career in chemical engineering, rest assured that the possibilities are endless, and your skills and expertise will be in high demand in various industries.

Continuing Education and Professional Development

In the ever-evolving field of Chemical Engineering, it is crucial for students to recognize the importance of continuing education and professional development. As a student, your educational journey does not end with a degree but rather sets the foundation for a lifelong learning process. This subchapter will delve into the significance of ongoing education and the opportunities available for professional growth in the field of Chemical Engineering.

Continuing education serves as a means for students to stay up-to-date with the latest advancements in their chosen field. The world of Chemical Engineering is constantly evolving, with new technologies, techniques, and processes being developed regularly. By engaging in continuing education programs, students can acquire the necessary knowledge and skills to adapt to these changes effectively.

There are numerous avenues for continuing education in Chemical Engineering. Professional organizations, such as the American Institute of Chemical Engineers (AIChE), offer workshops, conferences, and seminars focused on various aspects of the field. These events provide students with opportunities to learn from industry experts, network with professionals, and gain insights into the latest research and innovations.

Additionally, universities and colleges often offer postgraduate programs, such as Master's degrees and Doctoral programs, for students who wish to delve deeper into specific areas of Chemical Engineering. These advanced degrees can enhance your expertise, increase your marketability, and open doors to more specialized career opportunities.

Professional development is another crucial aspect of a student's journey in Chemical Engineering. It involves honing skills and acquiring knowledge beyond the traditional academic curriculum. This can be achieved through internships, cooperative education programs, and research projects. These experiences not only provide practical, hands-on experience but also allow students to develop critical problem-solving, communication, and teamwork skills.

Furthermore, students can engage in professional development by joining student chapters of professional organizations. These chapters often organize guest lectures, workshops, and industry visits, providing valuable exposure to the real-world applications of Chemical Engineering.

In conclusion, continuing education and professional development are integral components of a successful career in Chemical Engineering. By actively seeking opportunities for ongoing learning and growth, students can stay ahead of the curve, adapt to industry changes, and unlock their full potential. Remember, education does not end with a degree; it is a lifelong journey that will continually shape your career trajectory in this dynamic and exciting field.

www.ingramcontent.com/pod-product-compliance
Lightning Source LLC
Chambersburg PA
CBHW072034150726
47999CB00002B/894